Busick Harwood

A Synopsis of a Course of Lectures on Anatomy and Physiology

Busick Harwood

A Synopsis of a Course of Lectures on Anatomy and Physiology

ISBN/EAN: 9783337138899

Printed in Europe, USA, Canada, Australia, Japan

Cover: Foto ©berggeist007 / pixelio.de

More available books at **www.hansebooks.com**

T O

RICHARD FARMER, D.D.

MASTER of EMMANUEL COLLEGE

A N D

PRINCIPAL LIBRARIAN

IN THE UNIVERSITY OF CAMBRIDGE,

THE FOLLOWING

S Y N O P S I S

O F A

C O U R S E of L E C T U R E S

O N

Anatomy and Physiology,

IS MOST RESPECTFULLY INSCRIBED,

B Y

HIS MOST OBEDIENT HUMBLE SERVANT,

BUSICK HARWOOD.

EMMANUEL COLLEGE,
1st Feb. 1792.

ADVERTISEMENT.

HAVING found that the COMPENDIUM ANA-TOMICUM made ufe of by my predeceffor, was infufficient to anfwer all the purpofes of the more enlarged plan, which I have ventured to adopt in the delivery of anatomical lectures; I have been led to attempt the compofition of a Syllabus, which I hope will convey a more perfect idea of the fubjects I intend to enlarge upon in the enfuing courfe.

With regard to the order, in which I have chofen to arrange the different articles of this Syllabus, I have adhered as nearly to that in which they will be treated of at the lectures,

as

as the nature of the undertaking would admit. And to remedy the inconvenience, which might arife from any occafional breach in this order, I have avoided the divifion of it into feparate. lectures, and have prefixed numbers to each article, by which means they may be referred to at pleafure. The number of articles has alfo been reduced into as fmall a compafs as poffible, by omitting the enumeration of the Bones and Mufcles, a catalogue of which will be found at the end of the Syllabus.

Befides the more immediate purpofe for which the following pages were defigned, I am not without hopes that an attempt (which as far as I know is the firft,) to collect and arrange the principal facts, and difcoveries in Anatomy, may be of ufe to other Teachers of the art, who may poffibly think it worth while to extend and improve the plan which I now offer to the Public.

After the Anatomy of the human body is completed, it is likewife my intention to give

fome

fome feparate lectures on the ftructure of Animals. And in the divifion of the fubject, all
thofe Analogies which affift us in explaining
the ftructure and ufes of the Animal Œconomy,
will be particularly pointed out.

The articles which compofe this part of the
Syllabus, are comprifed under the general head
of *Comparative Anatomy*.

INTRODUCTION.

History of Anatomy.

1. **R**ISE, progress, and present state of the science. Of its general utility.

2. Of the various kinds of preparations made use of, in investigating the more obscure parts of the human frame.

3. Explanation of the instruments, and the manner of using them, for the purpose of preparing, and preserving, the different parts of an animal body.

4. Necessary cautions concerning the use of anatomical preparations.

A   5. Explica-

5. Explication of the general terms of Anatomy.

Of the Nature, and Properties of the
B L O O D.

6. Recent blood appears like an homogeneous fluid.
Of the means employed to diſcover its compoſition.

Of the ſeparate Parts of the Blood.

SERUM.

7. The properties and uſe of this fluid.
Saline particles contained in it.

CRASSAMENTUM.

8. Compoſed of two parts.
Separation of theſe parts by waſhing in water.

Particular

Particular Examination of each.

RED GLOBULES.

9. Suppofed caufe of their red colour.

Various opinions concerning their nature and formation.

Microfcopic obfervations, &c.

10. Theories of Lewenhoec, Hewfon and others.

COAGULABLE LYMPH.

11. Its peculiar properties.

Is the caufe of the fpontaneous feparation of the blood into two parts.

12. Theory of the formation, and re-generation of parts, by means of the *Coagulable Lymph.*

13. Morbid appearances of the blood.

14. Difference between arterial and venous blood.

A 2   15. Prieftly's

15. Prieftly's doctrine, and experiments.

16. Human blood compared with that of animals.

Hiftory of Transfufion.

17. Method of performing the operation.

18. Effects of *Transfufion* on various animals.

General Divifions of the Human Body.

19. Hippocrates's divifion into CONTINENTES, CONTENTÆ & IMPETUM FACIENTES.

Other divifions of the antients.

Of ANIMAL FIBRES.

20. Their ftructure.

Infinite divifibility of an animal fibre.

Various

Various kinds of fibres.

Their gradual increafe and elongation.

21. Obfervations on the preternatural growth of giants.

22. Of wounds in general.

23. Cicatrices how formed.

Of MEMBRANES.

24. General idea of their ftructure and various ufe.

Their different degrees of fenfibility, in a healthy, and in a morbid ftate.

Of B O N E S.

25. The beginning and progrefs of *Offification.*

Of the variety of this procefs in the flat, cylindric, and fpherical bones.

26. General ftructure of bones.
Cancelli how formed.

Remarks

Remarks on the growth of bones.

27. Chemical analyfis of bone.

28. Of the varieties in the fhape and fubftance of bones.

Of the comparative ftrength of bones.

29. Of their nerves and blood-veffels.

Exquifite fenfibility of bones in fome difeafes.

Evident effect of diet on the bones.

Of the bones of animals fed on Madder.

30. External conformation of the bones.

Apophyfes, Epiphyfes, &c.

Of the Periosteum.

31. Manner in which it is connected to the bones.

Texture and ufes of this membrane.

Of

Of the MARROW.

32. Its nature and ufe.

Increafe of its fenfibility in difeafe.

33. Of the *pori medullares.*

Remarks on the curious diftribution of thefe pores.

34. Difeafes of the bones.

35. General effects of the *Lues Venerea* upon them.

Exhibition of the different appearances produced in them by this
difeafe.

36. Remarks concerning the treatment
of fimple and compound fractures.

Callus how formed, &c.

37. Ufe of the bones in general.

Of CARTILAGES.

38. Their fituation and manifold ufes.

39. Difference of their ftructure.

40. Are claffed under three general
heads.

41. Of the Perichondrium, and its ufe.

Of Ligaments.

42. Their variety.

43. Of the ftructure, and ufe of ligaments in general.

44. Of capfular ligaments in particular.

45. Practical remarks concerning the treatment of wounds of the ligaments, and cartilages.

Of ARTICULATION.

46. Of the feveral fpecies of articulation, admitting different degrees of motion.

47. Explanation of each particular fpecies.

48. General account of the fynovial glands, and of the fluid fecreted by them, for the lubrication of the joints.

49. Practical

49. Practical remarks on the different kinds of luxations, and the modes of reducing them.

50. Of Anchylosis.

SKELETON.

51. Of the different kinds of Skeletons, and methods of preparing them.

52. General division of the Skeleton.

* *Bones of the Head, Trunk, and Extremities, separately considered.*

Of the CRANIUM.

53. Natural figure, size, and unequal thickness of the skull.

54. Of the tables of the cranium.

Of the diploë and its uses.

55. The skull composed of several bones.

B   56. *Particular

---

* See a Catalogue of the bones of the Human Skeleton, at the end of this Syllabus.

56. * Particular defcription of the *Su-tures*, by which thefe bones are conne&ed with each other.

57. Advantages arifing from this mode of connexion.

Sutures often obliterated.

Of their accidental varieties.

Of the *Offa Triquetra*.

58. Obfervations on the original con-formation of the fkull.

View of the external bafis of the fkull.

59. Of the various *proceffes* obfervable in this part of the cranium.

Their names, fituations, and ufes.

60. Obfervations on the general, and particular ufes of each procefs.

61. General view of the internal ca-vity of the cranium.

62. Of

* The principal Sutures are SUTURA { CORONALIS. SAGITTALIS. LAMBDOIDALIS, SQUAMOSA,

62. Of the various impreffions, pits, and furrows in this cavity.

Marks of the longitudinal, and lateral finus's, &c.

63. Divifion of its internal bafis.

64. Defcription and ufe of the *Procef-fes*, &c. in the cavity of the fkull.

65. Remarks on the attachment of the *Dura Mater*.

66. Of the different *foramina*, for the paffage of blood-veffels and nerves.

Obfervations on the ufe of each.

67. Remarks on the entrance of the *carotid* and *vertebral* arteries into the fkull.

68. Of fractures of the fkull.

Of concuffions.

Operation of the trepan, when neceffary and how performed.

Cautions to the furgeon in per-forming this operation.

69. Defcrip-

69. Defcription of the feparate bones of the cranium.

70. Obfervations on the form, ftruc-ture, and particular ufe of each.

Of the bones of the face.

71. Of their connexion with the proper bones of the fkull.

72. A particular defcription of each of thefe bones.

73. Of their general, and particular ufes.

74. Of the formation and ufe of the Lachrymal Duct.

75. Remarks on the manner of per-forming the operation for the *Fiftula Lachrymalis.*

76. Of Caries, and other difeafes of the bones of the face.

77. Of the lower Jaw.
Of its Articulation:

Of the alveolar proceſſes.

Of the abſorption of theſe proceſſes in old age.

78. Of the teeth in general.

bOf the ſtructure and form of the
different claſſes of teeth.

Of their enamel, and its uſes.

79. Obſervations on the paſſage of the
nerves into the teeth.

Of the original formation of the
teeth, &c.

80. Of their diſeaſes.

Remarks on the uſual method of
drawing teeth.

Of the SPINE.

81. Of the vertebræ in general.

Of their ſtructure.

82. Of the proceſſes of the vertebræ
and their ſeparate uſes.

83. Of the large canal for the tranſ-
miſſion of the ſpinal marrow.

Of

Of the lateral holes for the paffage of the nerves.

Of the Vertebræ of the Neck.

84. Of the ATLAS and EPISTROPHEUS.
Obfervations on their peculiar
form, and articulation.

85. Of the perforations of the cervi-
cal vertebræ, for the paffage of
the vertebral artery.

Of the Dorfal Vertebræ.

86. Of their fubftance, fize, &c.

87. Of their articulation with the
ribs.

Of the Lumbar Vertebræ.

88. Of their fituation and ftrength.

89 Of the peculiarities of the vertebræ of the back and loins.

90. Of the ligaments connecting the
vertebræ.

91. Of their intervening cartilages.

92. Of the incurvations of the fpine.

93. Of its mechanifm and ufes.

94. Deformities of the fpine.
    Their caufes, and methods of
    cure.

Of   the   P E L V I S.

Of the Os Sacrum.

95. Of the falfe vertebræ and holes
    for the paffage of nerves.

96. Of the Os Coccygis.
    Its ftructure and ufe.

Of the Os Innominatum.

97. Compofed of the *Ilium, Ifchium,*
    and *Pubis.*

98. Thefe bones feparately confi-
    dered.

99. Acetabulum how formed.

100. Of the Symphyfis Pubis.

101. Remarks on the ftructure, and
    different capacities, of the Pelvis,
    in the male and female Skeleton.

102. De-

102. Deformities of the Pelvis confidered as the caufe of difficult births.

Of the T H O R A X.

103. Of the Ribs in general.
Their divifion into true and falfe ribs.
Of their form, fituation, &c.

104. Of the cartilages of the ribs.
Obfervations on the ufe of thefe cartilages.

105. Of the deformities and difeafes of the ribs.
Remarks on the treatment, and confequences of fractured ribs.

Of the Sternum.

106. Of the feparate bones of the fternum.
Of the *xyphoid* cartilage.

107. Remarks

107. Remarks on the general ſtructure, and uſe of the thorax.

108. Obſervations on the motion of the ribs, and ſternum, in reſpiration.

Of the Upper Extremity.

109. Of the SCAPULA.
Its ſtructure, ſhape, proceſſes, &c.
Of its articulation with the clavicle, and with the humerus.

110. Of the CLAVICLE.
Its ſituation, ſtructure and uſes.

111. Of the HUMERUS.
Structure, proceſſes, and articulation of this bone.
Of the extent of its motion.

112. Of the treatment of fractures, and luxations, of the ſhoulder, and arm.

113. Of the RADIUS and ULNA.

The

The ſtructure, &c. of theſe bones, ſeparately examined.

Of their articulation with each other, with the humerus, and with the carpal bones.

114. Of their particular uſes, and the variety of their motions.

115. Of the interoſſeus ligament, and its uſes.

116. Of the CARPUS.

Of the eight bones of the carpus, their names, ſtructure, and different ſhapes.

Of their ſituation, and connexion with each other.

Of their articulation with the bones of the fore-arm, and metacarpus.

117. Obſervation on their motion and uſes.

118. Of the METACARPUS.

The bones of which it is compoſed, ſeparately conſidered.

Of

Of their articulations, &c.

119. Of the Bones of the Fingers and Thumb.

Of their ſtructure, articulations, &c.

120. Remarks on the number, ſituation, and uſes, of the *Oſſa Seſſamoidea.*

121. General Obſervations on the mechaniſm, and uſes, of the bones of the upper extremity.

Of the Lower Extremity.

122. Of the Os FEMORIS.

Its form and ſituation in general.

Particular deſcription of the ſtructure, and uſes, of the *trochanters, condyles,* and other remarkable parts of this bone.

Of its articulation with the *os innominatum.*

123. Obſer-

123. Obſervations on the motions of this joint.

124. Of luxations, fractures, &c. of the thigh bone, and methods of treating them.

125. Of the Tibia and Fibula.

General deſcription of their ſtructure and ſhape.

Separate examination of all the parts of theſe bones.

Of their connexion with each other.

Of the articulation of the tibia, with the os femoris.

126. Of the ſemilunar cartilages, crucial ligaments, &c.

127. Of the Petella, and its uſes.

128. Of luxations and fractures of the patella, and of the bones of the leg.

129. Of the Tarsus.

The

The names and ſtructure, of the ſeven bones which compoſe it.

Of their connexion with each other, and particular uſes.

130. Of the articulation of the *aſtragulus*, with the bones of the leg.

131. Of the METATURSUS and TOES.

Separate examination of theſe bones.

Of the difference between them, and thoſe of the metacarpus, and fingers.

132. Remarks on the mechaniſm of the foot, and the manner in which progreſſion is performed.

133. SKELETON of a FŒTUS examined, and compared with that of the adult ſubject.

134. Remarks on its ſtriking peculiarities, with regard to the number, ſhape,

fhape, connexion, and ufes of the bones.

135. Obfervations on the peculiar advantages of the cartilaginous ap-
pendages affixed to the bones of
a fœtus.

136. OF the general ſtructure, and various attachments, of muſcles.

137. Of ſimple, and compound muſcles.

138. Of the names of muſcles derived from their inſertion, connexion, form, ſituation, uſe, &c.

139. Muſcles voluntary, and involuntary.

140. Of the phænomena of muſcular action.

141. Hypotheſes concerning the immediate cauſe of muſcular motion.

142. Tendons how formed. Of their connexions, and uſes.

143. *Apon-*

143. *Aponeurofes.*
Their ftructure, and ufes.

* *Mufcles of the Abdomen.*

144. A particular defcription of their
ftructure, and mode of action.
Of their manifold ufe.
Great importance of the action
of thefe mufcles, to the animal
œconomy.

145. Of *Poupart's* ligament.

146. Particular defcription, and ufe
of the abdominal rings.

147. Remarks on the defcent of the
teftis.

148. Practical remarks on the different fpecies of *Herniæ.*

149. Obfervations on the manner of
performing the operation for
Bubonocele.

150. Of

* See a Lift of all the Mufcles at the end.

150. Of the treatment of wounds of the abdomen.

Mufcles of the Upper Extremity.

151. Of the mufcles which move the fcapula on the trunk.

152. Of thofe which move the os humeri on the fcapula.

153. Of thofe which move the bones of the fore-arm on the os humeri.

154. Of thofe which move the radius upon the ulna.

155. Of thofe which move the car- pus on the fore-arm.

156. Mufcles of the metacarpus and the fingers.

157. Of the general and particular ufes of the mufcles of the upper extremity.

Mufcles of the Lower Extremity.

158. Of the mufcles which move the thigh upon the pelvis.

D

159. Of

159. Of thofe which move the bones of the leg, upon the os femoris.

160. Of thofe which move the tarfus on the leg.

161. Mufcles which move the meta-tarfus, and the toes.

162. Remarks on the ftructure of each of the above mufcles.
Of their ufes.

Mufcles which move the Head on the Trunk.

163. Situation, ftructure, and ufe of each of thefe mufcles.

Mufcles of the Neck, Back, and Loins.

164. Their ftructure and fituation.
Of their general ufes.

165. *Of the Phyfiology of the Mufcles.*

Of

OF THE CONTENTS OF THE THORAX.

Of the Pleura.

166. Its situation and attachments.
    Of the structure, and uses of the pleura.

167. Of the *mediastinum,* and its uses.

168. Pathological remarks on the diseases of the pleura.
    Of the *hydrops pectoris.*

Of the Thymus.

169. Different states of this gland in the adult subject, and in the fœtus.
    Opinions concerning its use.

Of the Pericardium.

170. Its structure, &c.
    Of the fluid contained in it.

Of

Of the ufe of the pericardium.

Of the HEART.

171. Of the fituation of the heart.

Of its form, and general ftructure.

Remarks on the difpofition of its mufcular fibres.

172. Divifion of the heart, into auricles, and ventricles.

Of the fepta between the auricles and ventricles.

Of the *foramen ovale.*

Obfervations on its being fometimes found pervious in the adult fubject.

Of the Right Auricle.

173. Its form, fubftance, and fituation.

Of the opening of the two *venæ cavæ* into it.

Of

Of its communication with the right ventricle.

Of the Right Ventricle.

174. Its particular ſtructure, ſituation, and capacity.

175. The *columnæ carneæ* how formed.
Their particular uſes explained.

176. Of the *valvulæ tricuſpidales,* placed between the right ventricle and auricle.
Their ſtructure and uſe.

177. Of the communication of the right ventricle with the pulmonary artery.
Of the *valvulæ ſemilunares.*
Their uſe, &c.

Of the Left Auricle.

178. Its ſtructure, &c.
Of the entrance of the pulmonary veins into it.

179. Of its communication with the left ventricle.

Of the Left Ventricle.

180. Its fubftance, fituation, &c.

181. Of the *valvulæ mitrales*, placed between this ventricle, and its correfponding auricle.
Of the office of thefe valves, and their manner of acting.

182. Of the communication of the left ventricle with the AORTA.
Valves of the aorta.

183. Of the coronary veins and arteries.

184. Remarks on the different capacities, and ftrength of the two ventricles.

185. Obfervations on the involuntary action of the heart.

186. Of the ufe of the heart in general.

187. Of *polypi, aneurifm,* and other difeafes of the heart, and its veffels.

188. Of the ſtructure of arteries and veins, and their general uſe.

Of the *vaſa vaſorum.*

Of the anaſtomoſing branches of the blood veſſels.

Of the connexion between the extreme branches of arteries and veins.

189. Particular deſcription of the AORTA and VENA CAVA.

ꞌOF THE CIRCULATION OF THE BLOOD.

190. Of the ancient doctrines concerning it.

Of the diſcovery of it, by Harvey.

Particular Deſcription of the manner in which the Circulation is performed.

191. Of the SYSTOLE and DIASTOLE of the ventricles and auricles of the heart.

Of

Of their alternate action.

192. Of the paffage of the blood from the right ventricle to the lungs, by the pulmonary artery.

193. Of its return from the lungs to the left fide of the heart, by the pulmonary vein.

194. Of its exit from the left ventricle, and diftribution throughout the whole body, by the aorta and its branches.

195. Of its return to the right auricle of the heart, by the vena cava.

196. Experimental proofs of the reality of this mode of circulation.

197. Remarks on the quantity of blood, and the velocity with which it circulates in a healthy ftate.

198. Of the valves in the veins, their fituation and ufe.

199. Of the fyftole and diaftole of the arteries.

200. Ob-

200. Obſervations on the different kinds of pulſes, in various diſeaſes.

ˢOf the L U N G S.

201. Their ſituation, and figure.

Their general ſtructure, and diviſion into lobes.

202. Of the Bronchia, or air veſſels.

Their particular ſtructure.

Of their ramifications, and the termination of them in the *veſiculæ bronchiales*.

203. Of the bronchial arteries and veins.

204. Of the *Pulmonary Artery* and *Vein*.

Diſtribution of theſe veſſels throughout the ſubſtance of the lungs.

E

Of their peculiar office.

205. Of the bronchial glands, and their ufe.

206. Of the *trachea*, or *afpera arteria*.

Its ftructure, fituation, and ufe.

Of RESPIRATION.

Of the Diaphragm, and Mufcles employed in Refpiration.

207. Situation, figure, and attachments of the diaphragm.

Of its *crura*, and *centrum tendinofum*.

Of the perforations of the diaphragm for the paffage of the *œfophagus* and *vena cava*.

208. Remarks on the alternate contraction and relaxation, of this important mufcle.

Hypothefes concerning the caufe of this phœnomenon.

209. Of

209. Of the fituation, and office of the intercoftal mufcles.

210. Of the manner in which refpiration is performed.

Of expiration and *infpiration*.

Effects of thefe oppofite actions, on the pulmonary veffels.

211. Of the primary ufes of refpiration.

212. Of the changes produced in the blood, in its paffage through the lungs.

Various opinions of authors on this fubject.

Prieftly's theory and experiments.

213. Of the fecondary ufes of the organs of refpiration.

214. Obfervations on the caufe of hickuping, and other fpafmodic affections of the diaphragm.

E 2

215. Of

215. Of *peripneumony, afthma, phthi-
fis pulmonalis,* and other difeafes
of the lungs.

216. Of fuffocation.

Remarks on the methods, made
ufe of, for the recovery of
drowned perfons.

217. General remarks on the peculiarities of the circulation in a
fœtus.

*Of the general diſtribution of the
Blood-Veſſels throughout the
body.*

218. Of the courfe of the principal
arteries and veins.

Of their names, origin, and re-
fpective ufes.

219. The peculiarities of each pointed out, and the caufes of them
inveſtigated.

220. Prac-

220. Practical remarks on the treatment of *Popliteal Aneurism*.

221. Of PHLEBOTOMY.

Of the arteries, and veins of the arm, in particular.

Their situation, separately, and relatively confidered.

222. Of the courfe of the nerves, with refpect to the blood-veffels.

223. Of the manner in which bleeding in the arm is ufually performed.

General rules and cautions to be obferved.

224. Of *Aneurifm* occafioned by puncture of the artery in bloodletting.

Reflections on other dangerous confequences of unfkilful venæfection.

225. ARTERIOTOMY, how performed, and when neceffary.

Of the BRAIN.

Of the Meninges of the Brain.

226. Of the DURA MATER.
    Its ftructure and attachments.
227. Of the proceffes of the dura mater and their ufes.
    Of the longitudinal, and lateral finus's, *Torcular Herophili*, &c.
228. Of the TUNICA ARACHNOIDÆA.
229. Of the PIA MATER.
    Its ftructure and ufes.
230. kOf the circulation of the blood in the brain.

General Divifion of the Brain.

231. Of the CEREBRUM.
    Divifion of this part into hemifpheres and lobes.
232. Of the cortical fubftance.

233. Of

233. Of the medullary fubftance.

234. Of the *corpus callofum, fornix,* &c.

235. Of the *corpora ftriata, thalami nervorum opticorum,* &c.

236. Of the *tubercula quadrigemina,* and *pineal gland.*

237. Of the four ventricles of the brain, and their communication with each other.

238. Of the infundibulum, pituitary gland, and other remarkable parts of the brain.

Of the Cerebellum and Medulla Oblongata.

239. Of their fituation and ftructure. Of their connexion with each other, and with the cerebrum.

240. Of their peculiarities.

241. Conjectures concerning the ufes of the feveral parts of the brain.

242. Of

242. Of the ten pair of nerves which pass out of the ENCEPHALON.

243. A particular description of each, and general account of their distribution, and uses.

244. Various hypotheses concerning the uses, and functions of the brain, and nerves in general.

An account of the principal experiments which have been made on this subject.

Of the Spinal Marrow.

245. Its exit from the brain and passage through the vertebral canal.

246. Of the nerves arising from the spinal marrow, and the distribution of them throughout the body.

247. Of pains in the head.

248. Of *apoplexy, palsy,* &c.

249. Of nervous disorders in general.

Of

250. Their divifion into fimple, and compound.

251. Of their excretory ducts.

Difpute between Ruysch and Malpighi concerning the ftructure of glands.

252. Of glandular fecretion, and the opinions of various authors on that fubject.

Of morbid fecretions.

253. Opinions concerning the doctrine of tranfudation.

254. Of *Schirrus, Cancer*, and other difeafes of the glands.

Of the Salival Glands.

255. Of the fituation, and office, of the *parotid, maxillary*, and *fub-lingual* glands.

256. O

256. Of their excretory ducts, and the entrance of them into the mouth.

257. Disagreeable consequences of wounds in these parts.

258. Of the nature and use of the *saliva.*

Of the Mouth in general.

259. Of the lips and palate.

260. Of the *glandulæ lenticulares.*

261. Of the *os hyoides,* and tongue.

262. Of the papillæ of the tongue, and the sense of tasting.

263. Of the *velum pendulum palati,* and *uvula.*

Consequence of their destruction by the *Lues venerea,* and other diseases.

264. Of the *pharynx* and *æsophagus.* Their situation, structure, action, and uses.

265. Of

265. Of the muscles employed in mastication, deglutition, and formation of the voice.

Of their separate uses, and mode of action.

Of the Oesophagus.

266. Phœnomena attending the passage of the aliment, from the mouth to the stomach.

267. Of *angina* or quinsy.

268. Of the LARYNX, and the cartilages which compose it.

Of the *glottis* and *epiglottis.*

Their structure, and peculiar uses.

269. Remarks on the general effects produced by the ossification of these parts.

270. Of the formation of articulate sounds.

Of

Of the A B D O M E N.

271. Of the external form, and regions of the abdomen.

272. Of its internal cavity.

273. Of the *peritonæum.*
Its ſtructure, proceſſes, extenſibility, and uſes.

274. Practical obſervations on the cauſes, and cure of *aſcites,* and other diſeaſes of the abdomen.

Of the Abdominal Viſcera.

275. Of the ſtomach.
Its ſtructure, ſize, and natural ſituation.
Of the coats of the ſtomach, and their uſes.
Of the action of the ſtomach.
Of the nerves, and blood veſſels of the ſtomach.

Of

Of the *cardia* and *pylorus*.

276. Remarks on the entrance and exit of the food, by thefe orifices.

Of DIGESTION.

277. Examination of the principal hypothefes which have been formed to explain the nature of this procefs.

Objections to each of thefe.

278. Of the difcoveries lately made on this fubject.

279. Of the *fuccus gaftricus*.

Examination into the nature and properties of this fluid in men and other animals.

280. Experiments of REAUMUR, SPALLANZANI and others.

281. Why the ftomach itfelf is not acted upon by the folvent power of the fuccus gaftricus.

282. Of

282. Of hunger and thirst.

283. Of diet in general.

284. Of substances which promote, or retard digestion.

Remarks on the pernicious effects of excessive drinking.

285. Of the effects of various kinds of poisons taken into the stomach.

286. Of *cardialgia*, indigestion, and other diseases of this organ.

287. Of the causes and effects of vomiting.

Remarks on the use of emetics in general.

Of the PANCREAS.

288. Its structure, situation, &c.

Of its excretory duct, and the entrance of it into the *duodenum*.

289. Of

289. Of the nature and properties of the pancreatic juice.

°Of the S P L E E N.

290. Its ſtructure, and ſituation.

Of the blood veſſels and nerves of the ſpleen.

291. Conjectures concerning its uſe.

Theories of HEWSON and others.

Of animals who have been de-prived of this organ.

292. Of the diſeaſes of the ſpleen, and pancreas.

ᴘOf the O M E N T U M.

293. Its figure, ſituation, and ſtruc-ture.

Of its blood veſſels.

Of the fat contained in the cells of the omentum.

Of

294. Of the ufe of the omentum.

Of the L I V E R.

295. Form, fituation, and general ftructure of this vifcus.

Of the *capfula gliffoni*, and external covering of the liver.

Of its nerves and blood veffels.

296. Of the *vena portarum*.

Its origin, diftribution, and extraordinary office.

297. Of the branches of the *cava* returning the blood from the liver.

298. Of the hepatic artery.

Obfervations concerning its ufe.

299. Of the *pori bilarii*.

300. Their origin, and the termination of them in the *Hepatic Duct*.

301. Of the *Gall Bladder*.

Its form, fituation, and office.

302. General

302. General obfervations concerning thofe animals which have no gall bladder.

303. Of the *Cyftic Duct.*

Its union with the hepatic duct, forming the *ductus communis choledochus.*

304. Entrance of the ductus communis into the *duodenum.*

305. Of the ufe of the liver.

306. Of the difference between the *cyftic* and *hepatic* biles.

307. Examination into the nature, and ufes of each.

308. Obfervations on the mixture of the bile and pancreatic juice with the imperfect chyle.

309. General account of *Chylification.*

310. Of the difeafes of the liver.

Of calculi in the gall bladder, &c.

311. Of

311. Of the causes, symptoms, and cure of the *jaundice*.

Of the INTESTINES in general.

312. Division of the intestines into large and small.

    Of their general structure, situation and length.

    Of the coats of the intestines.

    Of the *succus intestinalis*.

313. Of the peristaltic motion of the intestines.

314. Of the operation of cathartic medicines, &c.

315. General description of the *Mesentery*.

Of the DUODENUM.

316. Its connexion with the stomach, &c.

317. Entrance of the *ductus communis,* and *ductus pancreaticus* into it.

Of

Of the *valvulæ conniventes*, their ftructure and ufe.

Of the JEJUNUM and ILEUM.

318. Their fituation, extent, &c.

Remarks on the decreafe of the valvulæ conniventes in the ileum.

319. Of the entrance of this inteftine into the *Colon*.

Of the valve of the colon.

Mechanifm and ufe of this valve.

Obfervations on the fatal confe-quences of difeafe in this part.

'Of the CŒCUM.

320. Its figuré, fituation, and fup-pofed ufes.

*Of the variety of its fhape and fize in different animals.

321. Of the *Appendicula vermiformis cæci*.

G 2

Con-

---

* See comp. anat.

(52)

Conjectures concerning its use.

Of the COLON.

322. Description of the course and circumvolutions of this intestine.

Of its ligamentary bands and other peculiarities.

Of the RECTUM.

323. Its structure and situation.

Of the *rugæ* formed by the internal coat of the rectum.

324. Of the *sphincter* and *levatores ani.*

325. Of the glandular *lacunæ* of the large intestines.

326. Remarks on the difference of the structure and uses of the large and small intestines.

327. Of diseases and wounds of the intestines.

328. General observations on the use of clysters.

Particular defcription of the Mefentery.

329. Its mode of connexion with the
inteftines.

> Of the blood veffels, nerves,
> &c. of the mefentery and in-
> teftines.

> Diftribution and ufe of thefe
> veffels.

Of the LACTEAL VESSELS.

330. Of the firft difcovery of them
by *Afellius*.

> Of the general ftructure of the
> *lacteals* and *lymphatics*.

> Defcription and ufe of their nu-
> merous valves.

331. Of the *ampullulæ* of the lacteals.

> Demonftration of the lacteals on
> the inteftines of an animal re-
> cently killed.

332. Of their paffage to the *thoracic
duct*.

333. Of the *Receptaculum Chyli.*

334. Of the *Thoracic Duct.*

    Of the structure and situation of this duct.

    Of its termination in the angle between the *jugular* and *sub-clavian* veins on the left side.

    Use of this duct.

335. Remarks upon its passage, and the manner in which the *chyle* and *lymph* are propelled through it.

336. Observations on the manner in which animals are nourished.

    'Of the *Lymphatics.*

337. General history of the absorbent system.

    Discovery of this system in birds, fishes, and amphibia.

    Remarks on the importance of this discovery.

338. Of the origin of the lymphatics.
Their univerfality, &c.
Of their termination in the tho-
racic duct.

339. *Modus operandi* of thefe veffels.
Antient opinions concerning abforption by the *meferaic* veins.
Refutation of this doctrine by
the experiments of HUNTER
and others.
The abforbent powers of thefe
veffels, how promoted, &c.

Of *Lymphatic Glands.*

340. Their number, names, fize, and
fituation in different parts of the
body.
Their ftructure explained.
Of the manner in which the
lymphatic veffels are connected with them.
Of the fluid fecreted by them.

Ufes

(56)

Ufes attributed to them.

341. Of difeafes of the lymphatic glands arifing from the abforption of venereal matter, &c.

342. Of *Scrophula*.

343. Of the abforption of morbid matter from abfceffes, ulcers, &c.

344. Practical remarks on the treatment of *hydropic complaints* in general.

345. Of the methods of difcovering and injecting the lymphatic veffels.

Of the K I D N E Y S.

346. Of the fituation, and general ftructure of the kidneys.

Of their nerves and blood veffels.

347. Of the cortical, and medullary fubftances of the kidneys.

Of the *papillæ* and their orifices.

Of

(57)

"Of the *infundibula* and pelvis of the kidneys.

348. Remarks on the quantity of blood fent to the kidneys.

349. Fœtal peculiarities of the kidneys.

Of the ufe of the kidneys.

350. Of the URETERS.

Their ftructure, office, and courfe.

Of their entrance into the bladder.

351. Of the GLANDULÆ RENALES.

Of the BLADDER.

352. Its fituation, ftructure, and ufe.

353. Evacuation of the urine, how performed.

Of the mufcles employed in this office.

354. Analyfis of urine.

H

Remarks

355. Remarks on the different appearances of this fluid, in various difeafes.

356. Of *Calculi* in the kidneys, and the fymptoms attending their paffage to the bladder.

357. Of *Nephrotomy*.

358. Of ulcers and other difeafes of the kidneys.

359. Of *Calculi* in the urinary bladder.

360. Analyfis of calculi.

361. Experiments of HALES, and others.

362. Nature and properties of one fpecies inveftigated by SCHEELE.

363. A different fpecies difcovered.

364. Obfervations on the various kinds of folvents made ufe of, for the removal of the ftone.

365. Of *Lithotomy*, and the methods of performing that operation.

366. Of *Diabetes, Dysury,* suppression of urine, &c.

Of the Male Organs of Generation.

367. Of the SCROTUM.
Its structure and uses.

Of the TESTES.

368. Coats of the testis.
Of its general structure.

369. Of the spermatic artery and vein.

370. Of the *tubuli testis.*
The *Epidydimis* how formed, &c.

371. Of the use of the testes, and separation of the semen.

372. Of the properties of the semen, and its agency in the generation of animals.

373. Theories, and experiments of LEWENHOEC and others.

374. Of

374. Of *Hydrocele*.

375. Of *Hernia humoralis*, and other diseases of the testicles.

376. Of the *Vasa deferentia*.

377. Remarks on the structure, and passage of the spermatic cord.

378. Observations concerning eunuchs.

379. Of the *Vesiculæ seminales*.
Their texture, situation, &c.
Their use disputed by J. HUNTER.
Observations concerning them.

380. Of the *Corpora cavernosa Penis*.
Of their cellular structure, &c.

381. Of the *Urethra*.
Of the corpus spongiosum urethræ.
Of the bulb of the urethra.
Of the membranous portion of the urethra.

382. Of

382. Of the situation, structure, and office of the *Prostate Gland.*

383. Of Cowper's glands.

384. Lacunæ of the urethra, &c.

385. Of the Glans Penis and Præputium.

386. Of the blood vessels, nerves, and lymphatics of these parts.

387. Of the muscles called *Erectores,* and *Acceleratores.*

388. Of the separate, and combined uses of the above-mentioned parts.

389. Of strictures, and other diseases of the urethra.

Of the introduction of the catheter.

390. Observations on the use of *Bougies.*

391. Of Gonorrhœa.

Its symptoms, and effects.

392. Ob-

392. Obfervations on the ufe of mercury in this difeafe.

393. Of the Lues Venerea.

394. Of the attack and progrefs of this difeafe.

395. Of its local and conftitutional effects.

396. Of the treatment of the lues venerea in its different ftages.
Confequences of neglect or ignorance in the attempts to cure this difeafe.

397. Of mercurial preparations in general.

398. Of the effects of mercury on the conftitution.

Of the Female Organs of Generation.

399. Of the external parts.
Of the *Clitoris, Hymen,* and *caruncula myrtiformes.*

400. Of the *Urethra.*

Its

Its fituation, and ftructure, compared with the Urethra of the
male fubject.

401. Of the *Vagina.*

Of the *rugæ,* and nervous *papillæ*
of the vagina.

402. Of the *Uterus.*

Its fubftance, form, and fituation.

403. Of the *os internum, cervix,* and
fundus uteri.

Of the ligaments of the uterus.

404. Of its blood veffels, &c.

405. Of the *Tubæ fallopianæ.*

Their fituation, and connexion
with the *ovaria.*

406. Of the *Ovaria.*

Their ftructure, fituation, &c.

407. Of the *Ovula* and *Corpora lutea.*

408. Of the particular functions, and
combined ufes of the above
parts.

409. Of

409. Of the *Menses*.

Of the cause of the menstrual flux.

Of its natural duration and periodical return.

Consequences of the irregularity, obstruction, or excess of this evacuation.

410. Of the disorders incident to women at the first appearance, and natural cessation of this discharge.

411. Remarks on the cause of periodical hæmorrhages.

412. Of dropsy of the ovarium.

413. Of the diseases of the Uterus, &c.

Of Impregnation and Conception.

414. Of the state of the Uterus after conception.

Of

Of falfe conceptions.

Obfervations on the production of monfters.

415. Of extra-uterine conceptions.

Of the FŒTUS IN UTERO.

416. Of the *Placenta.*

Its ftructure and fituation.

Of its connexion with the Uterus.

417. Of the *umbilical cord.*

418. Of the veffels which compofe it, and their peculiarities.

419. Of the *Chorion* and *Amnios.*

420. Of the *Liquor Amnii.*

421. *Of the *Allantois.*

422. Remarks on the mode of communication between the mother and the foetus.

423. Of

I

---

* See comparative anatomy.

423. Of the manner in which it is supplied with nourishment.

Various opinions of authors on this subject.

424. Of the progressive changes which the fœtus undergoes, during the usual term of pregnancy.

425. Of the situation of the fœtus in utero at different periods of gestation.

426. Particular description of the circulation of the blood in a fœtus.

Of the *ductus arteriosus, ductus venosus*, &c.

427. Of *Parturition*.

Symptoms of approaching labour.

428. Observations on the usual modes of delivery.

Of

(67)

Of the extraction of the *fecun-*
dines.

429. Of the caufes of difficult labour.

430. Symptoms of the child's being
dead, &c.

431. Of preternatural births, and the
inftruments employed for the
extraction of the fœtus.

432. Of the *Cæfarean operation.*

433. Of the difeafes incident to women during pregnancy.

434. Of the breafts, and the changes
which take place in them during
geftation, and after child birth.

435. Of the diforders fubfequent to
parturition, &c.

436. Of *Abortion.*

Its caufes and prevention.

437. Experiments generally inftituted
(in fufpicious cafes) to afcertain,

I 2

whether

whether the child has been ftill born, or the contrary.

438. Obfervations on the ufual appearances in cafes of fufpected violence.

439. Cautions to perfons who may be called upon to give evidence on fuch occafions.

Of the EAR.

440. Of the external parts of the ear, and its muscles.

Of the *Meatus auditorius externus.*

441. Of the *Glandulæ Ceruminosæ.*

442. Of the *Membrana Tympani.*

Its structure, situation, attachments, &c.

443. Of the *Officula Auditus*, and the small muscles attached to them.

444. Of the *Labyrinth* of the ear.

445. *Vestibulum*, *semicircular canals*, and *cochlea.*

Of the two *Fenestræ.*

Of the mastoid cells, &c.

446. Of the *Tuba Eustachiana.*

447. Of the auditory nerves, and their distribution.

Of the *Chorda Tympani.*

448. Of the feparate and combined ufes of all thefe parts.

449. Of the fenfe of hearing.

450. Of deafnefs, and its caufes.

451. Remarks on the cure of nervous deafnefs, by electricity.

452. Of pain in the ear, and other difeafes of that organ.

Of the N O S E.

453. Of the external parts of the nofe.

Of the bones of the nofe.

Of its mufcles and cartilages.

454. Of the *Septum Narium.*

455. Remarks on the peculiar form, fituation, and ufe of the *Offa fpongiofa.*

456. Of

---

* See Art. 71 to 76.

456. Of the posterior openings of the nose.

457. Description of the sinus's which communicate with its internal cavity, and their uses.

458. Of the *lachrymal sac*.

459. Of the *Membrana Schneideriana*, or Pituitaria.

 Structure, and extent of this membrane.

 Of the mucus secreted on its surface.

 Remarks on the uses of the pituitary membrane.

460. Of the *olfactory* nerves.

 Their passage through the cribriform lamella of the æthmoid bone, and distribution of them over the pituitary membrane.

461. Of the sense of smelling.

462. Of nasal hæmorrhages, *Ozæna, Polypus*, &c.

463. *Of the external parts of the eye.*

Of the structure and use of the eye-brows.

Of the eye-lids, &c.

464. Of the lachrymal gland, and its excretories.

465. Structure and uses of the *puncta lachrymalia, caruncula lachryma-lis,* and lachrymal duct.

Of the Globe of the Eye.

466. Of the *Tunica conjunctiva,* or *adnata.*

Of the *Tunica albuginea.*

467. Of the six muscles subservient to the eye.

Explanation of their structure, situation, and action.

Remarks

Remarks on the uniform motion
of the eyes.

Of the three proper coats of the eye.

468. General explanation of the manner in which the coats of the eye
are formed.

469. Of the *Sclerotica* and *Cornea.*

Particular examination of the
ftructure and ufes of this
coat.

Of its connexion with the *Choroides.*

470. Of the *Choroides* and *Uvea* or
Iris.

Peculiarities of this coat.

471. Of the manner in which the
Uvea is formed.

472. Of the *Ligamentum Ciliare.*

473. Of the *Pigmentum nigrum.*

474. Of the *Pupil.*

K   * Re-

*Remarks on the variety of its ſtructure in different animals.

475. Of the expanſion and contraction of the pupil.

Several phœnomena hence accounted for.

476. Of the *Retina* and *Optic Nerve.*

477. The entrance of the optic nerve into the orbit, deſcribed.

The opinions concerning the manner in which the optic nerves are conjoined.

478. Of the ſtructure and diſtribution of the *Retina.*

Of the three humours of the eye.

479. General remarks on the ſituation, tranſparency, &c. of theſe humours.

480. Of the Chambers of the eye.

481. Of the *Aqueous Humour.*

Of

Of its different degrees of tran-
fparency, at different periods.

Remarks on the effect of the
Jaundice upon this humour.

482. Of the evaporation of the aqueous humour; and the regeneration of it after it has been arti-
ficially evacuated.

483. Pathological remarks on the
caufe of *Hydropthalmia.*

484. Of the *Cryftalline Humour.*

Of the ftructure and fituation of
the Cryftalline Lens, and its
Capfule.

485. Of its fhape and denfity.

Obfervations on its want of vifi-
ble attachment, &c.

Different ftates of it at different
periods of life.

486. Of the *Vitreous Humour.*

Remarks on the quantity, and
denfity of this humour.

                                   487. Of

487. Of the cellular ſtructure of the
membrane which contains it.
Of the cavity for the lodgement
of the cryſtalline, &c.

Of VISION.

488. Of the refracting powers, &c.
of the different humours of the
eye.

489. Of the reſpective uſes of the three
coats of the eye.

490. Obſervations on the uſe of the
ligamentum ciliare.

491. Of the change produced in the
eye, that objects may appear dif-
tinct at different diſtances.

492. Of the manner in which the
pictures of objects are formed
upon the Retina.

493. Of the *punctum cœcum,* and
Mariotte's experiment to
prove the inſenſibility of the
Retina at that part.

494. Diſpute

494. Difpute betwixt Mariotte and Pequet concerning the feat of vifion.

495. Of the caufes and effects of the contraction and dilatation of the pupil.

Why we fee beft from a dark place to a light one, and *vice verfa*.

496. Obfervations on the ufe of the *pigmentum nigrum.*

Why fome people fee better in the dark than others, &c.

Several other phœnomena accounted for.

497. Remarks on the caufe of *Myopia,* or fhort fight.

498. Of *Prefbyopia,* or weak fight.

Obfervations on the ufe of glaffes to remedy the above imperfections.

499. Of *Cataract.*

Of

Of couching, and the manner of performing that operation.

500. Of the various caufes of *fquinting*, and its ufual remedies.

501. Of other difeafes of the eye, and the methods of treating them.

Of the TEGUMENTS of the Body.

502. Of the *Cuticula* or *Epidermis.*

Its ftructure and office.

Of its fpeedy regeneration.

503. Remarks on the operation of blifters.

504. Of the *rete mucofum.*

Its fituation and ufe.

505. Obfervations on the caufe of the black colour of Negroes, &c.

506. *Of the Cutis or Skin.*

Its ftructure in general.

507. Of the papillæ of the fkin.

508. Ge-

508. General remarks on the fenfe of FEELING.

509. Of infenfible tranfpiration.

510. Of the *glandulæ miliares*.

511. Of the perfpirable matter.
Of its fecretion, paffage, and efcape from the body.

512. Of the quantity of this evacuation.
Experiments of SANCTORIUS, and others.

513. Of the cutaneous Lymphatics and of abforption from the furface of the fkin.

514. Obfervations on the ufe of nutritive baths.

515. Of the *Membrana adipofa*, and reticular fubftance.
Their texture and ufes.

516. An enquiry into the nature and ufe of fat.

517. Obfervations on the treatment of *Emphyfema*.

518. Of

518. Of the Nails.

Their ſtructure, uſe, &c.

519. Of the Hairs.

Manner of their growth and re-
ceiving nouriſhment.

Of their general utility.

520. Of the *alopecia,* and *plica polo-
nica.*

521. Obſervations on the collections
of hair found in the ſtomachs of
animals, and in ſome abſceſſes,
&c.

522. Of the ſuppoſed integuments of
the antients.

523. RECAPITULATION of the principal phœnomena of the animal
œconomy, which have been
taken notice of in the preceding
lectures.

CONCLUSION.

COMPARATIVE ANATOMY.

INTRODUCTION.

1. OF the utility of comparing the
ſtructure of man with that of animals.

Of the uſe of comparative anatomy
in reading antient authors, and
its importance in the ſtudy of
Zoology, &c.

2. Obſervations on the comparative
ſize and ſtrength of animals.

3. Of the general analogy obſervable
in the ſtructure of animals.

4. Of the diviſion of animals into
claſſes.

L

Of

5. Of Quadrupeds.

6. Remarks on the peculiarities in the fkeletons of different quadrupeds.

7. Obfervations on the ftructure, and ufes of the tails of various animals.

8. Of the Panniculus Carnofus.

9. Of the fituation of the heart compared with that of the fame organ in the human fpecies.

10. Of the brain and nerves of quadrupeds.

11. Of the *rete mirabile,* or plexus of blood veffels on each fide of the *fella turcica.*

Reflections on the ufe of this plexus.

12. Of the Nose of quadrupeds.

Of its internal ftructure, and of the paffage, and diftribution of the olfactory nerves.

Re-

Remarks on the caufe of the acute fenfe of fmelling in various animals.

13. Of the EARS.

Of the variety in the fhape, fituation and ufes of the external
ear.

14. Of the EYE.

Its ftructure in different animals.

Of the *mufculus fufpenforius.*

15. Of the membrana nictitans.

16. Of the figure of the pupil in different animals.

Of its extreme dilatability in
fome animals, as in the cat,
&c.

17. Of the different colours of the
choroid coat, or *Tapetum,* in different animals.

Why certain animals are ena-
bled to fee with very little
light.

18. Of

18. Of the ſtructure of the Teeth in various animals.

Of the difference between the teeth of granivorous, and thoſe of carnivorous animals.

19. Of the want of the *Uvula* in quadrupeds, and the uſe of the muſcle attached to the *Glottis*.

ANATOMY OF A DOG.

20. Of the *Omentum*.

Remarks on the ſize and extent of the omentum in quadrupeds.

21. Of the chylopoietic viſcera.

Of the longitudinal direction of the *valvulæ conniventes*.

Structure of the inteſtines of this animal compared with that of the human inteſtines.

22. Of the digeſtion of carnivorous animals.

Re-

Remarks on the fize of the *Appendix Vermiformis.*

Of the Pancreas Afellii.

Of the bags found at the extremity of the rectum.

23. Of the *pancreas, fpleen,* and *liver* of this animal.

24. The kidneys and urinary bladder, compared with the fame parts in the human fubject.

25. The openings in the *abdomen* for the paffage of the teftes to the fcrotum, compared with the abdominal rings in men.

26. Of the male parts of generation.

Of the want of *veficulæ feminales* in the dog.

Remarks on the tedious copulation of this animal.

27. Of the female parts of generation.

28. Of the defect of fenfible perfpiration in this fpecies.

29. Remarks on the caufe and prevention of the *Rabies canina*.

Of the *Hydrophobia* and the attempts to cure this difeafe.

30. Of the peculiarities obfervable in the anatomy of a HORSE.

31. Of the courfe of the principal blood veffels, &c.

32. Remarks on fome of the moft common difeafes of horfes.

OF RUMINANT ANIMALS.

33. *Of the four ftomachs of ruminant animals, compared with the fingle ftomach of other quadrupeds.

Of

* All ruminant animals have more than one ftomach.

Of the manner in which rumina-
tion is performed.

Why ruminant animals require
lefs food than other granivorous quadrupeds, who have but
one ftomach.

34. Of the parts of generation of a
cow.

Of the *Uterus* and its *Cornua*.

35. Of the Fœtus.

36. Of the *Urachus* and *Allantois*, or
refervoir of urine peculiar to the
fœtus of quadrupeds.

Of Birds.

37. General remarks on this clafs of
animals.

38. Of the peculiarities in the fkeletons of birds.

39. Of the ftomach and inteftines of
the carnivorous, compared with
thofe of granivorous birds.

40. Of

40. Of the *fuccus gaftricus*, and the digeftive faculties of the carnivorous tribe.

42. Of the *ventriculus fuccenturiatus*, and gizzard.

43. Of the triturating power of the gizzard, and the manner in which their food is digefted.

44. Of the abforbent fyftem in birds.

45. Of the kidneys and paffage of the urinary fecretion.

46. Of the extent and attachment of the lungs.

Of their communication with the abdominal veficles, and the air cells in the bones of thefe animals.

47. Of the Diaphragm.

48. Of the brain compared with that of quadrupeds.

49. Of the olfactory nerves, and the organ of fmelling.

50. Of

50. Of the eye, and its peculiarities in this class of animals.

Of the *Marsupium nigrum,* or *Bourse noire.*

51. Of the organ of hearing in birds.

52. Of the organs of generation in the male.

53. Of the organs of generation in the female.

Of the vitellarium, infundibulum, &c.

54. Of the phœnomena attending the passage of the egg to the uterus.

55. Structure of an egg explained.

56. Of the progressive changes which the egg undergoes during incubation.

Of the nourishment of the fœtus of oviparous animals.

57. Of

M

57. *Of the secretion in the crops of breeding pidgeons for the nourishment of their young.

OF AMPHIBIOUS ANIMALS.

58. Of the heart and lungs of amphibia.

Of the peculiarities in the structure of these organs.

59. Of the transverse canals in the septum between the ventricles, exemplified in the heart of a turtle.

Of the use of these canals.

60. Why all the arteries proceed from the right ventricle.

61. Description of the circulation of the blood in this class of animals.

62. Ob-

* See a Differtation on this fubject, and alfo an account of the Air Cells in the Bones of Birds; in Obfervations on certain parts of the Animal Œconomy, lately publifhed by Mr. J. Hunter.

62. Obfervations on the *pulmo arbi-trarius* enjoyed by them.

63. General remarks on the ftructure of *ferpents*.

64. Of the teeth of ferpents, and their canal for the paffage of the poifonous fluid.

Of the refervoir in which this fluid is contained.

65. Of the general effects of wounds made by the teeth of venomous ferpents.

66. Obfervations on the treatment of perfons bit by this fpecies of ani-mals.

Of Fishes.

67. Remarks on the ftructure and ufe of the fins, tail, and other ex-ternal parts.

68. Of the fituation, and ftructure of the teeth of fifhes.

69. Of

69. Of the organs of digestion, and of the chylopoietic canal.

70. Of the swimming bladder and its use.

71. Of the size and structure of the liver.

Of the situation of the gall blad-der, and course of the hepatic and cystic ducts, &c.

72. Of the *intestinula cœca*, and their terminations.

73. Of the spleen, &c.

74. Description of the heart, and its vessels.

Of the single form of the heart, being composed of one auricle and one ventricle.

75. *Of the circulation of the blood in this class of animals.

Of

---

* See a Treatise on the Structure and Physiology of Fishes, published in 1785, by Alex. Monro, M. D. Professor of Physic, Anatomy, and Surgery, in the University of Edinburgh.

Of the paſſage of the blood from the ventricle of the heart to the gills, by the bronchial artery.

Of the union of the bronchial veins forming trunks, which perform the office of arteries, and convey the blood, (by their ramifications) all over the body.

Of the return of the blood to the heart by the *Venæ cavæ*.

76. Remarks on this mode of circulation, and on the uſe of the gills.

Obſervations on the peculiarities of the circulation in the *Sepia Loligo*, or cuttle fiſh.

77. Of the abſorbent ſyſtem in fiſhes.

78. Of the Brain and Nerves.

79. Of the organ of ſmell in fiſhes.

80. Of the Ear.

81. Its

81. Its ſtructure and ſituation in the cartilaginous and oſſeous fiſhes.

82. Obſervations on the faculty of hearing in water.

83. Of the eye.
Peculiarities of the cryſtalline, &c.

84. Of the *mucus duċts*, and the ſecretion of the liquor for the lubrication of the external ſurface of fiſhes.

85. Of the parts of generation in the cartilaginous and oſſeous fiſhes.

86. Of *Inſects, Vermes*, &c.
Conclusion.

*OS Frontis.
Os Parietale
Os Temporale
*Os Occipitis
*Os Ethmoidale
*Os Sphænoidale
Os Unguis
Os Nafi
Os Malæ
Os Maxillare
Os Spongiofum inferius
Os Palati
*Vomer
*Maxilla inferior

Dentes { Incifores / Canini / Molares

*Os Hyoides

Officula Auditus { Stapes / Incus / Malleus / Os Orbiculare

*Vertebræ { Colli feptem / Dorfi duodecim / Lumborum quinq.

*Os Sacrum
*Os Coccygis

Os Innominatum ex { Ilio / Ifchio / Pube

Coftæ { Veræ feptem. / Spuriæ quinque

*Sternum
Clavicula
Scapula
Os Humeri
Ulna
Radius

Offis Carpi { Scaphoides / Lunare / Cuneiforme / Pififorme / Trapezium / Trapezoides / Magnum / Unciforme

Offa Metacarpi
Offa Digitorum quindecim
Os Femoris
Tibia
Fibula
Patella

Offa Tarfi { Aftragalus / Os Calcis / Os Naviculare / Offa Cuneiformia / Os Cuboides

Offa Metatarfi quinque
Offa Digitorum Pedis quatuordecim

MUS-

Thofe bones which are fingle in the fkeleton, have no afterifk before them,—the reft are in pairs.

Abdominis
{
Obliquus externus.
Obliquus internus.
Tranfverfalis.
Rectus.
Pyramidalis.
}

Frontis et Occipitis.   Occipito-Frontalis.
Superciliorum.   Corrugator.

Genarum
{
Quadratus, five Platyfma Myoides.
Buccinator.
}

Palpebrarum
{
Orbicularis.
Aperiens Rectus.
}

Alarum Nafi
{
Levator.
Dilator.
Conftrictor five Tranfverfalis.
}

Labiorum
{
Elevator } Labii Superioris proprius
Depreffor }

Elevator } Labii Inferioris proprius
Depreffor }

Elevator } Communis.
Depreffor }

Zygomaticus major, et minor.
Orbicularis.
}

Auriculam
{
Attollens.
Retrahens.
}

Oculorum
{
Obliquus { Superior, five Trochlearis.
           Inferior.
Attollens.
Deprimens.
Abductor.
Adductor.
}

Auris Internæ
{
Externus.
Internus.
Obliquus.
Stapedæus.
}

Offis Hyoidis
{
Sterno-hyoidæus.
Coraco-hyoidæus.
Stylo-hyoidæus.
Genio-hyoidæus.
Milo-hyoidæus.
}

Linguæ {
Genio-gloſſus.
Stylo-gloſſus.
Baſio-chondro-cerato Gloſſus.
Linguales.

Uvulæ {
Gloſſio-Staphylinus.
Pterigo-Staphylinus.
Pharyngo-Staphylinus.
Levator, ſive Salpingo-Staphylinus.

Pharyngæus.
Oeſophagæus.

Laryngis {
Sterno-thyroidæus.
Hyo-thyroidæus.
Crico-thyroidæus.
Crico-Arytenoidæus { poſticus. lateralis.
Arytenoidæus major, et minor.
Thyro-Arytenoidæus.

Maxillæ Inferioris {
Temporalis.
Maſſeter.
Digaſtricus.
Pterigoidæus externus, et internus.

Thoracis {
Serratas Anticus major, et minor.
Subclavius.
Scalenus.
Triangularis Sterni.
DIAPHRAGMA.
Intercoſtales externi, et interni.
Serratus Poſticus ſuperior, et inferior.

Scapulæ {
Trapezius, ſive Cucullaris.
Rhomboides.
Levator, ſive Muſculus Patientiæ.

Capitis {
Splenius.
Complexus.
Rectus Poſterior, major, et minor.
Obliquus ſuperior, et inferior.
Sterno-maſtoidæus.
Trachelo-maſtoidæus.
Rectus Anterior { major. minor. lateralis.

Colli {
Longus.
Spinalis.
Semiſpinalis.
Tranſverſalis.
Inter-ſpinales.
Inter-tranſverſales.

Dorſi

Dorſi { Sacrolumbalis.
{ Longiſſimus.

Lumborum { Quadratus.
{ Sacer.
{ Pſoas Parvus.

Coccygæus.

Humeri {
Pectoralis.
Deltoides.
Supra-ſpinatus.
Infra-ſpinatus.
Teres minor, et major.
Latiſſimus Dorſi.
Coraco-brachialis.
Subſcapularis.

Humeri {
Biceps Flexor.
Brachiæus internus.
Biceps extenſor, ſive Gemellus.
Brachiæus externus.
Anconæus.

Volæ Manus.   Palmarus longus, et brevis.

Radii { Supinator longus, et brevis.
{ Pronator Teres, et quadratus.

Carpi { Flexor, Radialis, et Ulnaris.
{ Extenſor, Radialis, et Ulnaris.

Digitorum Communes {
Extenſor.
Flexor { Sublimis, ſive perforatus.
{ Profundus, ſive perforans.
Lumbricales.
Interoſſei.

Indicis { Indicator.
{ Adductor proprius, et communis.

Digiti minimi {
Abductor.
Primi internodii flexor.
Extenſor.

Pollicis {
Flexor internodii primi, ſecundi, et tertii ſive longus.
Extenſor internodii primi, ſecundi, et tertii.
Abductor.
Adductor, proprius et communis.

Flexor Oſſis Metacarpi, minimum Digitum ſuſtinentis.

Femoris {
Pſoas magnus.
Iliacus internus.
Pectineus.
Glutæus magnus, medius et minimus.
Triceps Extenſor.
Iliacus Externus, ſive pyriformis.
Gemelli.
Obturator externus, et internus ſive Marſupialis.
Quadratus.

Cruris

Cruris {
Membranofus.
Sartorius.
Gracilis.
Bracilis.
Biceps Flexor.
Seminervofus
Semimembranofus.
Rectus.
Vaftus externus, et internus.
Crureus.
Poplitæus.

Tarfi {
Tibialis Anticus.
Gaftrocnemius, ubi *Tendo Achillis.*
Plantaris.
Soleus.
Peroneus longus et brevis.
Tibialis Pofticus.

Digitorum Pedis {
Extenfor longus, et brevis.
Flexor Perforatus, et Perforans.
Lumbricales.
Interoffei.

Pollicis Pedis {
Extenfor longus, et brevis.
Flexor longus, et brevis.
Abductor.
Adductor.

Tranfverfalis Pedis.

Digiti Minimi {
Flexor Proprius.
Abductor.

Penis {
Erectores.
Acceleratores Urinæ.
Tranfverfalis.

Clitoridis, Erector.

Vaginæ Sphincter.

Ani {
Levatores.
Sphincter.

F I N I S.